21ST Century Skills INNOVATION LIBRARY

Exploring the Internet of Things

Smart Homes

Martin Gitlin

CHERRY LAKE PRESS

Published in the United States of America by Cherry Lake Publishing
Ann Arbor, Michigan
www.cherrylakepublishing.com

Reading Adviser: Marla Conn, MS, Ed., Literacy specialist, Read-Ability, Inc.

Photo Credits: © Andrey Suslov/Shutterstock, cover, 1; © metamorworks/Shutterstock, 5; © Tullio Saba/Flickr, 7; © adventtr/iStock, 8; © Proxima Studio/Shutterstock, 9; © Joni Hanebutt/Shutterstock, 10; © SoNelly/Shutterstock, 13; © Chris Allan/Shutterstock, 15; © Iglenas/Shutterstock, 16; © Casezy idea/Shutterstock, 18; © Sergey Klopotov/Dreamstime.com, 19; © okoland/Shutterstock, 21; © Konstantin Shadrin/Shutterstock, 23; © Aaron Yoo/Flickr, 24; © designmilk/Flickr, 27; © mentatdgt/Shutterstock, 29

Graphic Element Credits: © Ohn Mar/Shutterstock.com, back cover, multiple interior pages; © Dmitrieva Katerina/Shutterstock.com, back cover, multiple interior pages; © advent/Shutterstock.com, back cover, front cover, multiple interior pages; © Visual Generation/Shutterstock.com, multiple interior pages; © anfisa focusova/Shutterstock.com, front cover, multiple interior pages; © Babich Alexander/Shutterstock.com, back cover, front cover, multiple interior pages

Library of Congress Cataloging-in-Publication Data has been filed and is available at catalog.loc.gov.

Printed in the United States of America
Corporate Graphics

Table of Contents

Introduction

People yearn for peace of mind. They want to feel safe in their homes. Most also desire easy living. And that means getting things done without lifting a finger.

The Internet of Things (IoT) is turning that dream into reality. The technology uses **sensors** to connect everyday appliances and devices to each other and to the internet. IoT makes life easier. IoT makes it possible to turn up the heat in your home before you get there. Or check the camera at your front door for suspicious sounds.

Using technology to bring life to lifeless objects was always seen as futuristic. Home safety meant physically locking doors and hoping for the best. Those wanting to turn on a light had to get home and do it themselves—after stumbling around in the dark to find the light switch. But not anymore, thanks to IoT-connected home devices. Devices like the Google Nest allow you to manage your home

The Internet of Things is under the umbrella of the Internet of Everything (IoE).

remotely, and ensure that devices are turned off and not wasting energy.

Smart homes are the new reality. They provide information that helps homeowners make more informed decisions. IoT technology allows for a safer, more comfortable, and more convenient home. The smart home industry is quickly growing. It is producing more and more devices that allow people to live better and easier lives. According to experts, the smart home industry will potentially be valued over $144 billion by 2025.

Most people spend a good amount of time at home. Some want to make that time more productive. Others seek to make it more relaxing. Many hope to achieve both depending on their mood. IoT technology makes it possible to attain those goals.

CHAPTER ONE

Devices and Uses

IoT can bring ease to home life. It can even achieve that when people are away. One example is someone driving home from work. Despite being tired, this person still needs to cook dinner and feed the dog. IoT technology can help. Its sensors could command it to spill food into the dog dish and set the oven for 350 degrees Fahrenheit (177 degrees Celsius).

Perhaps it is freezing outside. People can save money by turning down the heat before leaving every morning. But by the afternoon, the house is cold. However, an IoT-connected **thermostat** can raise the heat level at a specific time. This means people can save money *and* return to a warm, cozy home.

IoT-connected devices can handle many ordinary tasks. They can turn lights on as people enter a room and switch them off after they leave. They can raise the window shades in the morning and pull them down at night.

The concept of a "smart home" was first developed as early as the late 1950s!

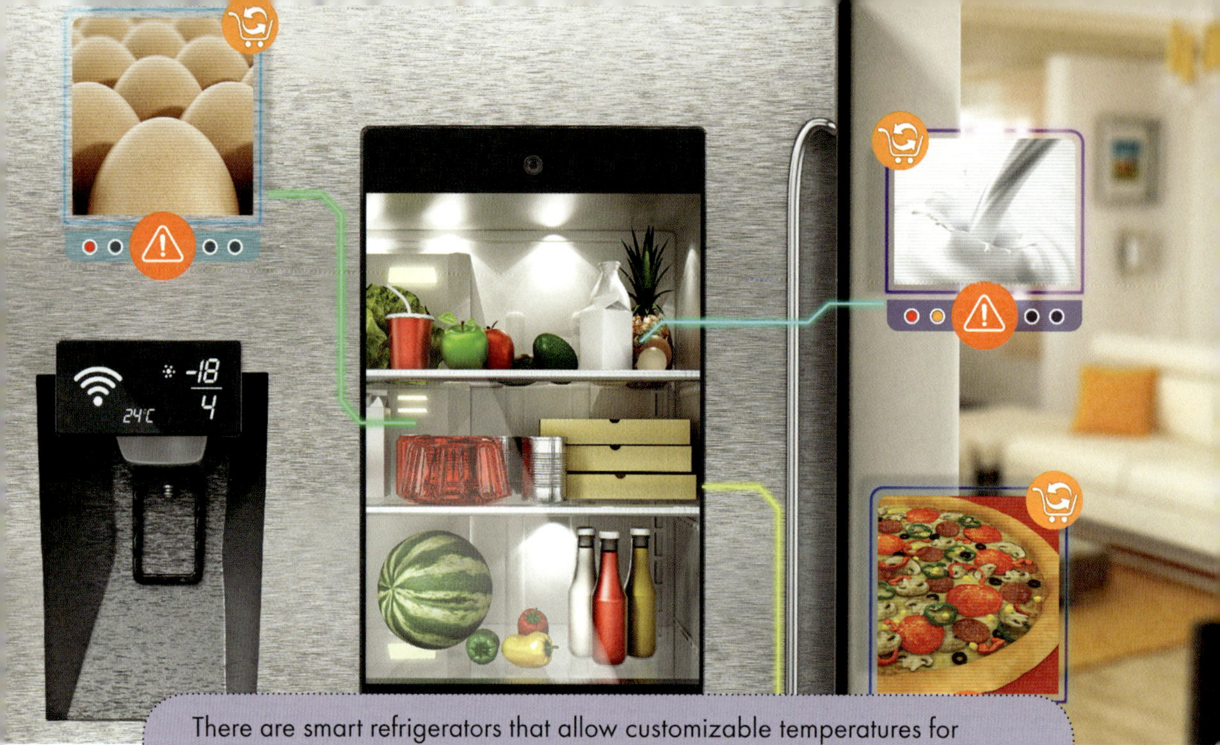

There are smart refrigerators that allow customizable temperatures for different drawers.

But IoT technology extends beyond simple tasks. For example, a smart refrigerator does more than keep food cold. It can send recipes to smartphones. It can tell people how to cook a chicken as soon as it's removed from the refrigerator. Did you forget what you need to buy at the grocery store? Smart refrigerators also allow people to check what's in their fridge through their phone. It can also help you reduce the time spent cooking dinner. In 2016, the USDA found that American adults spend 37 minutes on average cooking and cleaning their meals. Smart technology could reduce that time and make the cooking and cleanup process more efficient.

An IoT-connected coffee maker can have a cup made and poured minutes after a person wakes up. A toaster **embedded**

Some smart coffee makers and juicers even connect to your smartphone and even your smart watch!

Home assistants, like Echo or Google Home, can help kids with reading and math and even time management.

with a sensor can have hot toast ready to be buttered in the morning. A smart washing machine can turn on only when enough clothes have filled it up.

IoT-connected devices also provide **dental hygiene** *and* entertainment for kids. For instance, Grush is a smart toothbrush that turns brushing teeth into a fun interactive game. Kids can also turn on the television without lifting a remote. Smart TVs feature voice recognition. Anyone can command it to play any channel at any time. Much like a smartphone, parents can control what their child watches on a smart TV.

IoT technology can save adults and kids from doing a few chores. One example is sweeping and vacuuming the floors. Robot vacuum cleaners can do that job without human help. Connected lawnmowers can cut the front and back lawns. Plants can be watered through connected timers. The system can sense freezing pipes and turn off the water to prevent flooding.

Smart homes have the potential to bring safety and convenience to people, even when they are thousands of miles away. One example is automatic IoT door locks, in case you forgot to lock the door when you left home. IoT technology allows people to **monitor** their empty homes from anywhere. Smart motion and camera sensors tell the difference between friendly visitors, such as dog sitters, and burglars. They can notify the police of attempted break-ins.

Smart homes are certainly convenient. But do they have any drawbacks?

A SMART ... EGG TRAY?

IoT devices can connect to just about anything—even an egg tray! The Egg Minder allows people to view the number of eggs in a tray through a mobile device **app**. The app notifies egg lovers whether they should buy another dozen. But the device does more than that. It can also detect the health of an egg. It sends alerts when an egg has gone bad or is close to going bad. It also knows which eggs are the freshest. The result is that people can pick out the older eggs to cook first.

Researchers predict that by 2022 people in North America will spend about $63 billion in smart home services and systems.

CHAPTER TWO

Benefits and Concerns

Perhaps the biggest problem with smart homes is not knowing the best use of smart home devices. Many people worry that the technology is too complex.

That is why smart home companies have tried to simplify the process. They have worked to make home **automation** easier while proving its benefits. **Manufacturers** of IoT devices seek to convince potential users that they make life easier. For instance, smart homes remember to do things people might forget or don't have time to do. Gardens can be watered. Floors can be vacuumed. The lights and air-conditioning can be turned on and off when not in use.

More and more people are converting their homes into smart homes. In North America, about 22 million households had connected by the end of 2016. That number is expected to reach 63 million by 2022 in North America alone.

According to a study, smart thermostats can help people save about 25 percent on their cooling and heating bill.

About 76 percent of smart home devices are controlled by smartphones. This might spell disaster if a phone were hacked into.

While this is great news, this also presents a security problem. It was once obvious when a break-in had occurred. A window had been smashed or a door broken open. But IoT technology has made break-ins harder to detect. A smart door can be unlocked remotely. Cameras can be **hacked**. Homes can be secretly monitored. Cisco, a cybersecurity company, had to block 20 billion threats a day in 2019 in order to keep their customers' data secure.

Smart home devices operate remotely online and through mobile apps. Users must protect themselves against hackers. They must also protect all their connected devices and appliances. A 2016 study showed that most users were concerned about hackers. Criminals capable of hacking smart devices could turn off lights and alarms and unlock doors. That would leave a home defenseless against a physical break-in.

Those worried about hacking had reasons to be. Smart home devices have already been targeted by hackers. In 2016, hackers took over hundreds of thousands IoT devices and used them to send fake internet traffic to target websites. Their goal was to crash the sites. The hacking wreaked havoc with internet connections in parts of North America and Europe. Such problems continued. Hackers made big news in 2019 by taking over in-home smart cameras and scaring people in their bedrooms and living rooms.

Another IoT issue is cost. Smart home technology might be rejected by those seeking to save money. Turning traditional homes into IoT homes can be expensive. On average, it costs about $5,000 dollars. Some old appliances and devices can be integrated into smart homes. But some IoT technology requires buying new smart appliances. New switches and plugs could be needed to turn lighting, climate control, and security systems into smart devices. Others argue that IoT saves money on heating and air-conditioning bills. Being able to control when they are turned on and off saves energy costs.

In order to protect your smart home devices from hackers, experts suggest using a secure Wi-Fi router.

The benefits or drawbacks of smart homes are **subjective**. IoT technology is embraced by those who feel it brings convenience and peace of mind. It is rejected by those who prefer self-control of their homes and don't feel the benefits outweigh the costs.

Smart home companies will keep working to convince the skeptics. They will keep trying to advance their technology while making it easier to use. Their success or failure will help determine the future of IoT in the home.

There are smart robots, like the CANBOT, that can learn the habits and personality of people to better teach them!

KEEPING TRACK OF ANIMALS

Keeping tabs on a dog can be difficult. Dogs move around quickly and sometimes quietly. And few people have the time or desire to follow them around to ensure they are not getting into trouble. Smart collars can solve that problem. They use **GPS** to track dogs and other animals. Some smart collars also allow users to create a zone in which their pets can play. Owners are alerted through text or email when their pets have strayed outside that area. Farmers have also benefited from smart collar technology. They can place them around the necks of their farm animals. That allows them to keep track of their horses and cows remotely.

There's a smart water fountain for cats! It can track drinking habits and even behavioral changes.

CHAPTER THREE

The Future of Smart Homes

One does not need a wild imagination to think about a future when nearly all American homes will be simplified by IoT technology. People will be turning lights on and off with their voices or through an app. Robots will be vacuuming floors and helping cook dinner. Connected lightbulbs will let owners check on their pets while they are at work.

Smart home manufacturers will stay busy trying to **personalize** the technology. Among those companies is Crestron. It has worked on software that tracks homeowner habits. The result will be lights that know when to turn on and off. Or televisions that switch on when a favorite show begins. Or a song that plays automatically in the morning.

Even robotic furniture is possible. Beds and desks could change places during the day and move back at night. A smart dinner table could fold in the morning and expand in time for dinner.

According to a study, about 72 percent of people in the United States want their smart home devices to have voice-control capabilities.

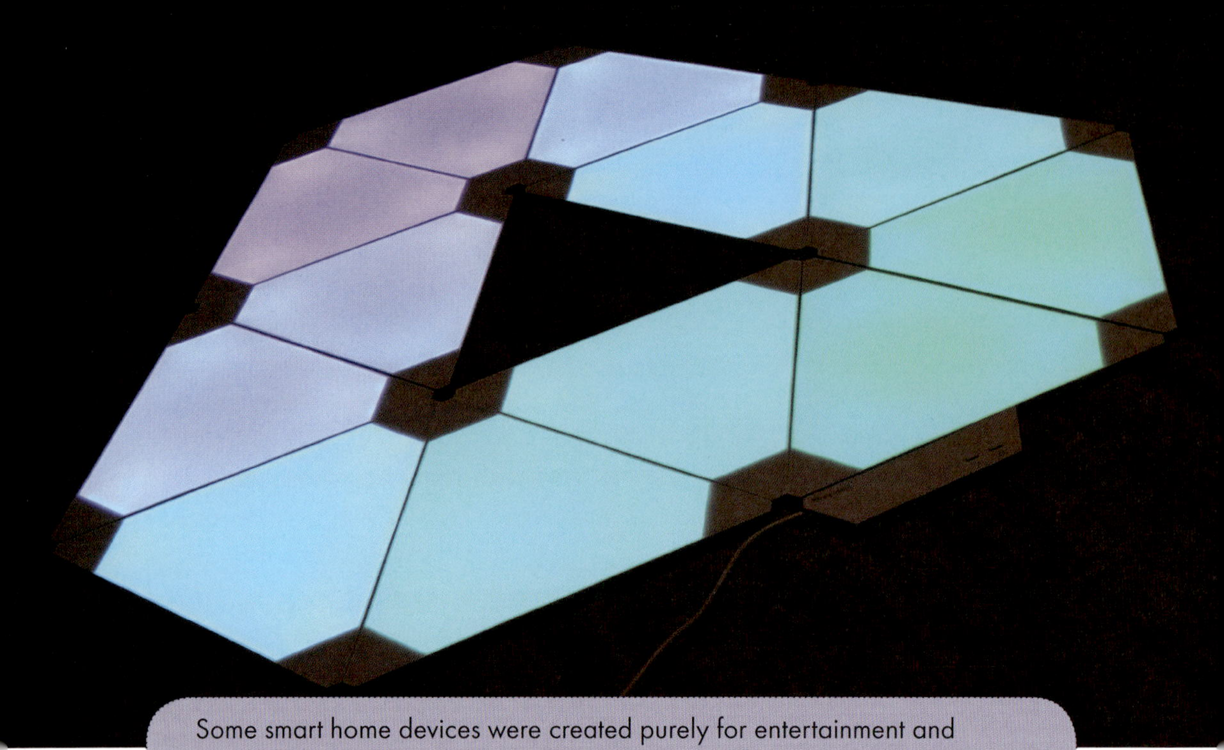

Some smart home devices were created purely for entertainment and decoration, like the customizable smart light panels.

IoT technology could move beyond fitness and activity trackers to help monitor health. Cameras and sensors in refrigerators can warn people against reaching for sugary sodas. They can suggest a nutritious piece of fruit instead.

No two people are alike. Smart home manufacturers understand that. So they will try to customize IoT systems to individual wants and needs in the future. Some people might prefer IoT devices in the kitchen. They enjoy having their morning coffee brewed for them. They want food they put in the oven a little earlier to be cooked automatically while they drive home. But perhaps these same people enjoy doing their own yard work and housecleaning. They want to water their own lawns and vacuum their own floors.

The growing number of smart home gadgets will allow people to choose which ones they want. More companies will specialize in connecting to particular IoT devices and machines. Research conducted by the International Data Corporation claimed that **consumers** will spend $195 billion on devices driven by IoT in 2023.

Smart device manufacturers will work to make use even easier. One way to achieve that goal is through technology that requires only voice commands rather than **interface** programming. That way, smart home benefits would take far less work. Another way to achieve that goal is by technology that can complete several tasks with one request. One IoT hub taking voice commands could be instructed to turn up the heat, set the oven for 400 degrees, and play a favorite song all at 6 p.m. And it would be done. There would be no need to program the thermostat, oven, and music device individually.

IoT technology promises to grow in many areas of human existence. But how it is accepted in homes around the world remains to be seen. We are progressing into a period known as the Fourth Industrial Revolution, in which IoT technology will likely change everyday habits.

ONE SMART TRASH CAN

The Internet of Things has inspired countless products that help with everyday tasks—including throwing out the trash! Garbi is a smart device that connects to a trash can. It helps people sort recyclables, compost, and actual trash. It does this while also creating a grocery list that can be easily purchased through connected sites like Instacart, Amazon, and Costco. Garbi uses Computer Vision, the same technology used in automated cars. The smart trash can takes a photo of the item being discarded. It analyzes the item using the internet to determine where to repurchase the item and whether it should be recycled, composted, or simply thrown away.

Some smart trash cans also change the trash bag automatically!

Creativity and Innovation

IoT devices in the home will continue to grow. Product developers work to think of new ideas. Those ideas are based on the needs and desires of potential buyers.

Smart home devices must help owners in at least one of three areas. They should make home life safer, easier, or more comfortable. The result is that they can be helpful in a home even when a person is not there. It takes great imagination to come up with new ideas.

Think About It

Research current smart home devices. How would you make them better? What new devices can you come up with?

There are smart mirrors that takes photos of the outfits you wear. Because it syncs with your online calender, it has the ability to notify if you've already worn a particular outfit when meeting someone again.

Learn More

Books

Macauley, David. *The Way Things Work Now*. Boston, MA: Houghton Mifflin Harcourt Publishing, 2016.

Rooney, Anne. *You Wouldn't Want to Live Without the Internet!* New York, NY: Franklin Watts, 2016.

Swanson, Jennifer. *National Geographic Kids—Everything Robotics: All the Robotic Photos, Facts, and Fun*. Washington, DC: National Geographic Society, 2016.

Websites

Kinooze—The Internet of Things
http://kinooze.com/the-internet-of-things
Learn more about the world of IoT.

Wonderopolis—What Makes a Home "Smart"?
https://www.wonderopolis.org/wonder/how-does-a-house-become-a-smart-home
Discover more about smart homes and the benefits and challenges in living in one.

Glossary

app (AP) computer application designed for mobile systems and devices

automation (aw-tuh-MAY-shuhn) use of machines rather than people to do work

consumers (kuhn-SOO-murz) people who buy things

dental (DEN-tuhl) relating to teeth

embedded (em-BED-id) set solidly in another object

GPS (JEE PEE ES) a system that uses satellite signals to determine a user's location and give directions; stands for global positioning system

hacked (HAKT) gained illegal access to

hygiene (HYE-jeen) practices you do to stay clean in order to stay healthy

interface (IN-tur-fase) program used to communicate between different computer systems

manufacturers (man-yuh-FAK-chur-urz) companies that make things in large amounts, such as in a factory

monitor (MAH-nih-tur) to check on or keep track of something

personalize (PUR-suh-nuhl-ize) to make or create something with individual tastes or needs in mind

sensors (SEN-surz) detection devices that respond by transmitting a signal

subjective (suhb-JEK-tiv) personal opinion or belief

thermostat (THUR-muh-stat) device that controls indoor temperature

Index